# Truck Carrier Partner 2.0.13 Tool:
# Quick Start Guide
# 2013 Data Year - United States Version

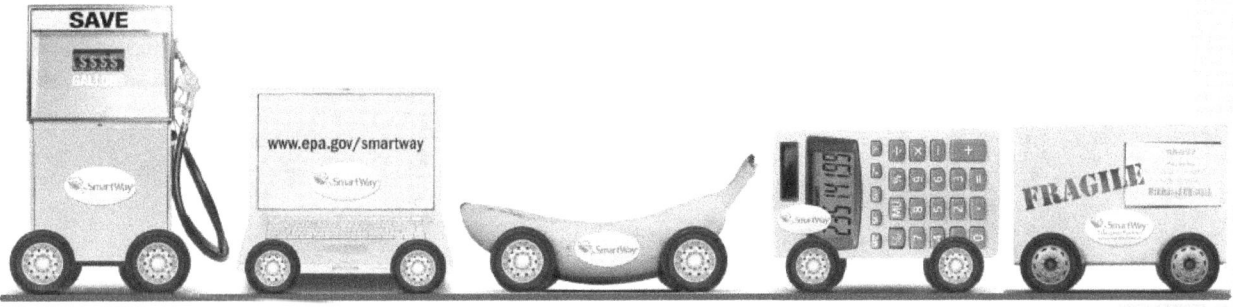

United States
Environmental Protection
Agency

# Truck Carrier Partner 2.0.13 Tool:
# Quick Start Guide
# 2013 Data Year - United States Version

Transportation and Climate Division
Office of Transportation and Air Quality
U.S. Environmental Protection Agency

EPA United States
Environmental Protection
Agency

Office of Transportation and Air Quality
EPA-420-B-14-001
January 2014

# Table of Contents

# Overview

In this guide you will learn about:

1) SmartWay Basic Information

2) Joining SmartWay Transport Partnership as a truck carrier

3) Understanding the details of the SmartWay partnership agreement

4) Meeting software/hardware requirements for participating in the program

5) Gathering the data necessary for participation in SmartWay.

**Please review this guide carefully BEFORE attempting to gather your company data, or enter data into the Truck Carrier Tool.** Understanding the basics of the program will simplify your SmartWay experience.

WARNING:

***Before beginning, use this chart to make sure you are choosing the right tool for your operations! ***

## Is this the right SmartWay tool for me?

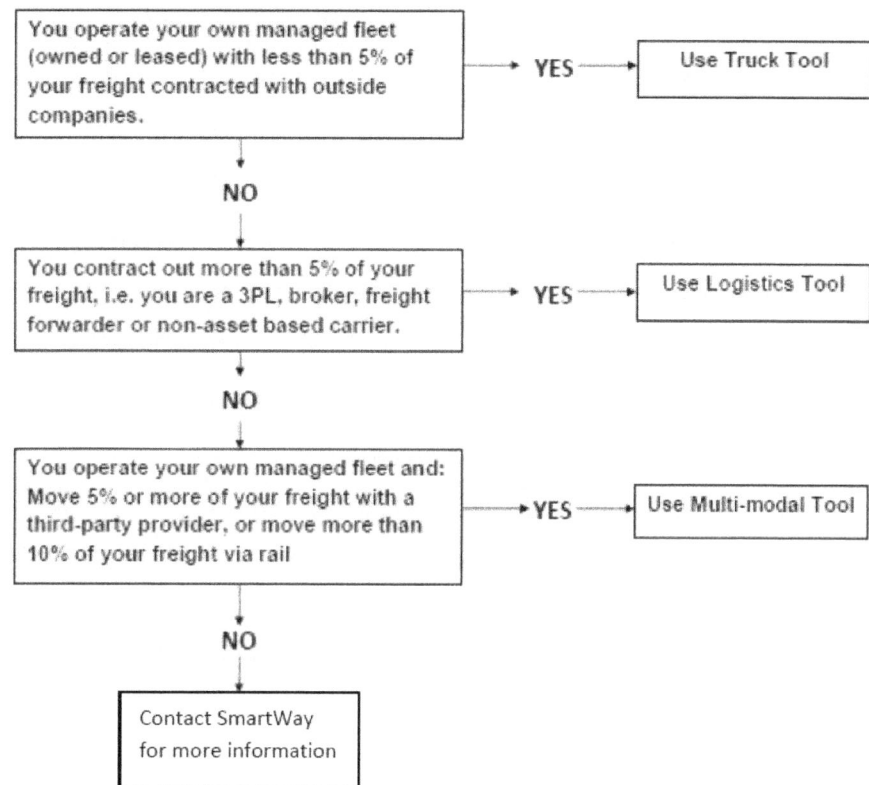

If none of the above statements is applicable, contact EPA SmartWay at 734-214-4647 for assistance.

## Section 1—Basic Information for Truck Carriers

This section covers frequently asked questions and essential information about the SmartWay Transport Partnership and how truck carriers can participate.

### WHAT IS THE SMARTWAY TRANSPORT PARTNERSHIP?

Launched in 2003, the SmartWay Transport Partnership is a public/private collaboration between the EPA and the freight industry to improve fuel efficiency, increase environmental performance, and encourage supply chain sustainability.

Five types of freight transport companies can join SmartWay.

- Freight shippers
- Logistics companies (including 3PLs/4PLs[1])
- Truck carriers
- Rail carriers
- Multi-modal carriers

Companies join the SmartWay Transport Partnership by submitting a Partner Tool to SmartWay. The SmartWay Tools (1) assess freight operations; (2) calculate fuel consumption and carbon footprints; and (3) track fuel-efficiency and emission reductions. SmartWay Tools must be submitted each year for the company to remain a Partner in good standing.

SmartWay ranks Partners' efficiency and environmental performance and recognizes superior performance through the SmartWay Excellence Awards.

### WHY DO TRUCK CARRIERS JOIN THE SMARTWAY TRANSPORT PARTNERSHIP?

The SmartWay Transport Partnership provides truck carriers with ways to reduce the environmental impact of their freight operations and address costs. Designed with and for the freight sector, the SmartWay Transport Partnership delivers solutions to marketplace needs and challenges. With access to the latest in EPA-tested technologies and peer-provided success stories, carriers that join the SmartWay Transport Partnership can gain a better understanding of their environmental footprint and assert their corporate leadership.

Additionally, SmartWay Partners are associated with an internationally recognized and respected brand that symbolizes cleaner, more efficient transportation choices.

---

[1] 3PLs/4PLs Third party logistics/fourth-party logistics companies.

## HOW DO TRUCK CARRIERS JOIN THE SMARTWAY TRANSPORT PARTNERSHIP?

Truck Carriers join SmartWay by submitting a SmartWay Truck Carrier Tool (hereafter known as the "Truck Carrier Tool.")

Truck Carriers that submit tools that are approved by EPA are known as "SmartWay Truck Carrier Partners."

When a truck carrier submits a Truck Carrier Tool to EPA, they agree to the requirements stipulated in the SmartWay Truck Carrier Partnership Agreement--notably, that they will measure and report the emissions performance of their company _annually_ and provide supporting documentation to EPA upon request.

All SmartWay Truck Carrier Partners agree to complete and submit the SmartWay Truck Carrier Tool to:

- define company composition
- characterize company activity
- individually benchmark fleets
- track annual changes in performance.

Upon approval of a Truck Carrier Tool submission, truck carriers will be identified as SmartWay Truck Carriers Partners on EPA's website on the SmartWay Partner List and in a database used to identify companies that meet SmartWay's annual requirements.


## HOW DO I JOIN IF MY PARENT COMPANY HAS MULTIPLE FLEETS?

Companies that join the SmartWay Partnership should include all of their fleets and trucks in their submission.  If a company wishes to list multiple fleets in the Truck Carrier Tool, they should list these fleets as their customers can hire them.  Internal fleets invisible to a customer should not be listed separately.  Companies will be listed at the Company level in the SmartWay Partner list on the SmartWay website, and each individually defined fleet will appear as a separate entity in the SmartWay Carrier Data file that customers use to identify which fleets they do business with in the Shipper and/or Logistics Tools.

*SmartWay highly recommends developing your list of fleets before beginning your data entry process.  Any fleet that a shipper or logistics company could hire directly should be listed as a separate fleet in your Truck Carrier Tool submission.*

*The best strategy is to have a clear idea of how to define your companies before filling out the Tool.*

## WHAT DATA DO I NEED TO GATHER TO COMPLETE THE TRUCK CARRIER TOOL?

To participate in SmartWay, truck carriers need to gather the following essential information to complete the Truck Carrier Tool:

- The official company name EXACTLY as you would like it presented on the EPA website
- Company contact information
- Contact details for your Primary Contact
- Contact details for an Executive and/or Other Contact(s) (cannot be the same as the Primary Contact)
- Split between US/Canada operations
- Quarterly IFTA statements[2] (for activity data) for the reporting calendar year
- Fleet details for all fleets you control:
  - SCACs, MCNs, or DOT number information
  - Total inventory of vehicles in your fleet(s), sorted by vehicle class and engine model year, body type, and operational category for the reporting calendar year
  - Total miles, revenue miles and empty miles
  - Total diesel, biodiesel and/or other fuel use by class
  - Reefer fuel use by class (if applicable)
  - Average payload, average capacity volume, and percent capacity utilization by class
  - Average idle hours per truck
  - Use of particulate matter control equipment by truck class and engine model year (if applicable)
- Data sources for all data to be entered
- SmartWay ID number (if this is not your first tool submission)

This data must be provided for all of your company's fleets. This data reflects the amount of freight carried by each carrier, the distance that freight is carried, and the fuel consumed to carry the freight.

---

[2] If applicable – for Class 7, 8a and 8b trucks only.

## WHAT IS INCLUDED IN THE SMARTWAY TRUCK CARRIER PARTNERSHIP AGREEMENT?

To join the SmartWay Transport Partnership as a truck carrier, you must agree to the language on the "Partnership Agreement for Truck Carriers." When you begin working within the Tool, you will be asked to check a box stating that you agree to the terms of the Partnership Agreement. **This agreement must be renewed annually.**

Please review this language with the appropriate personnel within your organization before completing or submitting a Tool to EPA.

### Partnership Agreement for Carriers

**With this agreement, your company joins EPA's SmartWay Transport Partnership and commits to:**

1. Measure and report to EPA on an annual basis the environmental performance of your fleet(s) using EPA's SmartWay Truck tool. (Existing fleets must report the 12 months of data for the prior year ending December 31. Newly formed companies require a minimum of 3 months of operational data.)
2. Have performance results posted on the EPA SmartWay website/database.
3. Agree to submit supporting documentation to EPA for any data used to complete this Tool and agree to EPA audit of this data upon request by EPA.

**In return, EPA commits to:**

1. Promote company participation in the Partnership by posting Partner names on the EPA SmartWay Website and in related educational, promotional, and media materials. EPA will obtain express written consent from the Partner before using the Partner's name other than in the context of increasing public awareness of its participation as described here.
2. Provide companies with industry-wide performance benchmark data as this data becomes available to EPA.[3]
3. Assist Partners in achieving emission and fuel usage reduction goals (subject to Federal Government Appropriations).

**General Terms**

1. If the Partner or EPA defaults upon this agreement at any point, the agreement shall be considered null and void.
2. Either party can terminate the agreement at any time without prior notification or penalties or any further obligation.
3. EPA agrees not to comment publicly regarding the withdrawal of specific partners.
4. EPA reserves the right to suspend or revoke Partner status for any Partner that fails to accomplish the specific actions to which it committed in the SmartWay Transport Partnership Agreement and subsequent annual Agreements.
5. The Partner agrees that it will not claim or imply that its participation in the SmartWay Transport Partnership constitutes EPA approval or endorsement of anything other than the Partner's commitment to the program. The Partner will not make statements or imply that EPA endorses the purchase or sale of the Partner's products and services or the views of the organization.
6. Submittal of this SmartWay Truck Tool annually constitutes agreement to all terms in this Partnership Agreement. No separate agreement need be submitted.

---

[3] Individual corporate data will be treated as sensitive business information.

## WHAT SOFTWARE AND HARDWARE IS REQUIRED FOR COMPLETING THE SMARTWAY TRUCK CARRIER PARTNER TOOL?

The Truck Carrier Tool was designed in "Microsoft Excel Forms." Completing the Truck Carrier Tool requires the following software and hardware:

- A 2003 or later version of Microsoft Excel

- Excel security level set at Medium or lower

- A PC running Windows XP or newer operating system, or a Mac that is running the Windows XP operating system (The Tool does not currently work using the Mac operating system)

- A minimum of 10 megabytes of free disk space. More disk space may be required based on the number of companies you define in your Tool

- Adequate memory (RAM) to run Microsoft Office

- A monitor resolution of at least 1,024 x 768

Please check with the user guides for your computer, online support, or your company's IT department to make sure your system is set up to use the Truck Carrier Tool.

We encourage you to make sure that you virus software is up to date, and scan your PC before putting data in the Truck Carrier Tool.

## Section 2— Overview of Data Collection Requirements

This section will explore the data needed for completing the required sections of the Truck Carrier Tool. **The Truck Carrier Tool Data Entry and Troubleshooting Guide** explains more about the structure of the Tool and the data entry process; this guide will focus primarily on the essentials for completing the Tool.

### DATA REQUIREMENTS FOR INTRODUCTORY SCREENS

There are four screens which orient you to the Tool, the Introduction, the Partnership Agreement, Tool selection guidance, and data collection needs. These are general information screens; however, note that you MUST click the box indicating that you agree to the terms of the Partnership Agreement before moving on to the next screen.

### DATA REQUIREMENTS FOR PROVIDING US/CANADA OPERATIONS INFORMATION

To begin, you must specify the percentage of your fleet(s) that is licensed in the United States vs. the percentage that is licensed in Canada. If you operate exclusively in one country or the other, simply enter 100% for the appropriate country.

Next, enter the percent of your total fuel consumed that is attributable to United States-licensed vehicles and to Canadian-licensed vehicles. Then specify the data source used to determine your fuel consumption allocation. If your data source is not specified you may select "Other" and provide a text description of this source.

Finally, if you have operations in both countries, you must indicate if these vehicles use the same fuel mix in both countries. If the fuel mix is different, you must describe the difference in the types and proportions of fuels used in the two countries in the text box provided in the Tool.

### DATA REQUIREMENTS FOR ESTABLISHING YOUR DATA COLLECTION YEAR

Before beginning your data collection, identify the last calendar year for which you have full annual (12 months) data. This means that you have data from January of the calendar year through December of the same year. If you are submitting for the first time and do not have a full year of operational data, please collect _**a minimum of three months' data**_ for input into the SmartWay Tool. In your next update year, you will be required to submit a full year's data.

## DATA REQUIREMENTS FOR SECTION 1: SPECIFY OFFICIAL PARTNER NAME

Your Partner Name is the official name that your customers would recognize for your company—in other words, the name someone hiring you would look for.

You must specify you company's official Partner Name, exactly as you want it to appear on the SmartWay website.

For example, if you enter:

- ABC Company
- ABC Company, Inc.
- ABC COMPANY LLC

Your company will be listed **_exactly_** as you've entered above.  Therefore, it is important to pay special attention to proper capitalization, abbreviations, annotations, and punctuation.

## DATA REQUIREMENTS FOR SECTION 2: ENTER CONTACT INFORMATION

The SmartWay Tool asks for:

- **General company information** such as location, web address, phone number, etc.

- **A Primary Contact**[4] for any questions about your company's participation and Tool submissions

- **An Executive Contact**[5] for participation in awards and recognition events

- **Additional contacts (optional):** Additional contacts may include anyone who is not the primary contact but may be involved with SmartWay (e.g., press/media contact, fleet manager, etc.).

Note that you MUST have at least two contacts listed in the contact information section of the Tool, and the Primary and Executive Contacts must be different.  SmartWay recommends developing an internal succession plan to make sure that your Truck Carrier Tool submission schedule is maintained, in the event that a Primary Contact is reassigned, or leaves the company for any reason.

---

[4] The Primary Contact is the individual designated by the Executive Contact to directly interface with SmartWay regarding specific tasks involved in the timely submission of the Tool.  The Primary Contact is responsible for coordinating the assembly of information to complete/update company data; completing and updating the Tool itself; maintaining direct communication with SmartWay; and keeping interested parties within the company apprised of relevant developments with SmartWay.)  NOTE: To ensure that emails from SmartWay/EPA are not blocked, new primary contacts may need to add SmartWay/EPA to their preferred list of trusted sources.)

[5] The Executive Contact is the company executive who is responsible for agreeing to the requirements in the SmartWay Partnership Agreement, overseeing the Primary Contact (as appropriate), and ensuring the timely submission of the Tool to SmartWay. The Executive Contact also represents the company at awards/recognition events.  This person should be a Vice President or higher-level representative for the company.

## DATA REQUIREMENTS FOR SECTION 3: CHARACTERIZE YOUR FLEETS

The Truck Carrier Tool tracks fleet-level operations. Most carriers should create one fleet i.e., one line.

However, if you have multiple fleets that can be hired by customers individually, you should create multiple fleet records in the Tool and characterize their operations individually. You should not include internal company fleet definitions or designations—only separate fleets as they would be identified and hired by your customers.

There are four screens in the **Fleet Characterization** section of the Tool:

- **Identify Fleets:** For each fleet that a customer can hire, you will be asked to establish names for each of your fleets. Each fleet name will begin with your Partner Name, and will be include a fleet identifier. Use a fleet identifier that is recognizable by your customers.

- **Fleet Details:** You will need to confirm control of the fleet. **"Control" means that you operate/route the vehicles, regardless of ownership status, and you must confirm that you have at least 95% control of the fleet to include it in the SmartWay Truck Carrier Tool.** You will also be asked to provide SCACS, MCN, or DOT numbers (for fleet identification), define the fleet as "for-hire" or "private," and identify a point of contact for the fleet.

- **Operation Categories:** You will need to provide a reasonable estimate of the fleet's split between truckload, less-than-truckload, drayage, package delivery, and expedited operations.

- **Body Types:** You will need to provide a reasonable estimate of the fleet's split between the following body types:

    - Dry van
    - Refrigerated (Reefer)[6]
    - Flatbed
    - Tanker
    - Intermodal chassis containers (pooled and owned)
    - Heavy/Bulk hauler
    - Auto carriers
    - Moving
    - Utility[7]
    - Special hauler (e.g., Hopper, Livestock, and other specialized carriers)

---

[6] If you specify reefer body types in your fleet you must also provide your estimated reefer fuel use in the Activity section of the Tool.

[7] The Utility category encompasses class 2b to 8b vehicles that do not carry typical commercial freight. Examples include garbage, recycle, service, work, dump, landscape, cement, bucket, boom, ambulance, armored, fire, farm, wrecker and other similar trucks. Because these trucks do not carry traditional freight payload, the user should self-define their payloads so as to make the emissions per payload efficiency useful to the user. SmartWay will not use the emissions per payload results for the utility category. Users may experience yellow or red warning labels on the Activity screen due to the unique nature of Utility "payload." In the case of red alerts, simply input text defining your special conditions in the required text boxes that appear.

## DATA REQUIREMENTS FOR SECTION 4: FLEET DATA

In Section 4 of the Truck Carrier Tool, you will enter **detailed activity and fuel consumption data** for each of the fleets you identified and characterized in Section 3.

### General Information Screen Requirements

This screen asks you to provide information for six key indicators, including fuel types used.

**The first section asks you to determine your short-haul split vs. long-haul split.** This requires you to estimate the percentage of your fleet's operations that are short haul or long haul. A short haul is defined as any haul less than 200 miles; a long haul is defined as any haul in excess of 200 miles. Percentages should be reasonable estimates.

**Next, you must select the fuel types found in this fleet.** This requires you to define all fuels the fleet used from a list of:

    a. **Diesel/biodiesel:** petroleum diesel and/or biodiesel made from any renewable feedstock

    b. **Gasoline/Ethanol:** conventional and reformulated gasoline, including blends of 10% ethanol(E10) and 85% ethanol (E85)

    c. **Liquefied Petroleum Gas (LPG):** also know as LP Gas, Liquid Propane Gas, and propane

    d. **Liquefied Natural Gas (LNG)**

    e. **Compressed Natural Gas (CNG)**

    f. **Electric:** vehicles exclusively using battery electric power

    g. **Hybrid:** vehicles with hybrid electric or hydraulic hybrid powertrains, using either gasoline or diesel

**NOTE:** Once you check these boxes, the appropriate fuel type tab (along the top of the screen next to the General Information tab) will become active. **You will need to complete additional screens for each fuel type you select.**

**You will be asked to determine whether you will need to enter data for Particulate Matter Reduction.** Check the Particulate Matter Reduction box only if you have truck engines that are model year 2006 or earlier and are equipped with diesel retrofit particulate matter control devices (i.e., diesel oxidation catalysts (DOCs), particulate filters, or closed crankcase ventilation (CCV)).

**Next you will be asked to determine what is known as your "Cube-Out" Percentage.** Your "cube out" percentage is the percentage of truckloads using 100% of their available cargo capacity while remaining within allowable weight limits. **You must provide a reasonable estimate.**

**Next you will go to the View/Select Commodities button to view a list of potential shipment types** and select all of the ones that you move with this fleet.

**You also have the option of choosing to participate in SmartWay's Port Dray Program.** Appendix A to the Truck Carrier Tool Data Entry and Troubleshooting Guide provides the data entry requirements for participation in this program, as well as details regarding Dray Program Scoring.

## Fuel-Specific Sections: Engine Model Year and Class, and Activity Information Screens

This section looks at truck classes and engine model years. Under each fuel-type tab (Diesel, Gasoline, LPG, LNG, CNG, Electric, and Hybrid) there are as many as four screens requiring data inputs:

1. the Engine Model Year & Class[8] screen

2. the Activity Information screen

3. the PM Reduction screen will appear for diesel vehicles <u>if</u> you checked the **Particulate Matter Reduction** box on the General Information screen

4. the Port Dray Program screen will also appear for diesel vehicles <u>if</u> you checked the **Port Dray Program** box on the General Information screen.[9]

**To complete these screens, you will need to know**

- the number of trucks in each fleet split out by engine model year and truck class.

- total miles by vehicle class

- revenue miles by vehicle class

- empty miles by vehicle class

- gallons of fuel used by vehicle class (or kWhrs for electric trucks)

- payload information by class and body type (if available; if not, use the Payload Calculator defaults)

- average capacity volume by body type and truck class for straight trucks (if available; if not, use the Volume Calculator defaults) or by trailer/container type for combination trucks

- road type/speed category splits by truck class

- average idling hours (short duration and extended) per truck for all classes

- the number of trucks by class and engine model year for which you are using particulate matter retrofit control devices

---

[8] For electric trucks, this screen is known as the "Motor Model Year and Class Screen".
[9] The Port Dray Program box will only appear if your fleet is predominantly defined as Dray – i.e., defined as greater than or equal to 75% Dray on the Operations Screen under Fleet Characterization.

## Section 3—Step by Step Instructions for Data Entry

### *DOWNLOADING THE SMARTWAY TRUCK CARRIER TOOL*

To download the Truck Carrier Tool, visit: http://epa.gov/smartway/partnership/trucks.htm . Save the Tool in a folder on your hard drive; this folder will house copies of your data and future updates.

### *HOW TO SET SECURITY LEVELS FOR THE SMARTWAY TOOLS*

In some cases, Microsoft Excel will ask you to adjust your security settings before opening the Tool. Instructions should appear on your screen *if* you need to change your security settings before running the Tool.  The instructions differ depending upon what version of Excel you use (Excel 2003, 2007 or 2010).

If you need additional assistance with your security settings, download the guidance document labeled "Truck Carrier Partner 2.0.13 Tool: Data Entry and Troubleshooting Guide.2013 Data Year—United States Version" from  http://epa.gov/smartway/partnership/trucks.htm  and review the screen-by-screen guidance in Part 1: Downloading and Setting Up the Tool.

## STEPS FOR ENTERING PARTNER NAME

1. Type your Partner Name EXACTLY as you would like it to appear on the SmartWay website in the field as indicated.

2. Proceed to Section 2 to enter contact information.

## STEPS FOR ENTERING CONTACT INFORMATION:

1. **Enter the Partner Information details** in Section 1 of the Contact Information screen.

2. **Enter Primary Contact details** in Section 2.   If the Primary Contact's address is the same as the company, you can select the 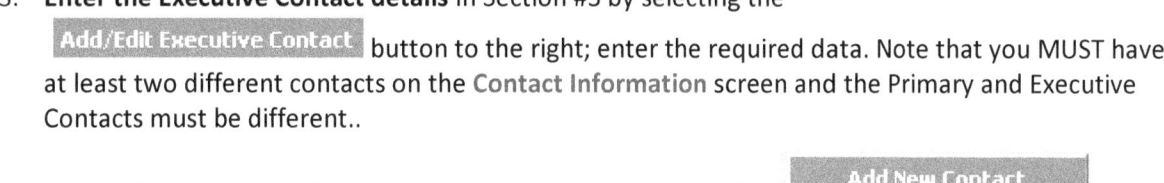 button to automatically fill in the address section of this record.

3. **Enter the Executive Contact details** in Section #3 by selecting the 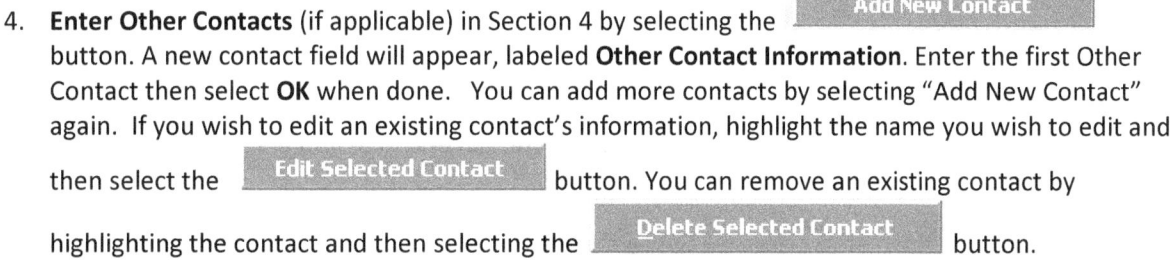 button to the right; enter the required data. Note that you MUST have at least two different contacts on the Contact Information screen and the Primary and Executive Contacts must be different..

4. **Enter Other Contacts** (if applicable) in Section 4 by selecting the ![Add New Contact] button. A new contact field will appear, labeled **Other Contact Information**. Enter the first Other Contact then select **OK** when done.   You can add more contacts by selecting "Add New Contact" again.  If you wish to edit an existing contact's information, highlight the name you wish to edit and then select the ![Edit Selected Contact] button. You can remove an existing contact by highlighting the contact and then selecting the ![Delete Selected Contact] button.

5. Select ![VALIDATE SCREEN] at the bottom of the screen.  If any information is missing, a dialogue box will appear informing you what additional information is required.

6. When finished, select the ![HOME] button to return to the Home screen and proceed to Section 3.

## STEPS FOR COMPLETING "IDENTIFY FLEETS" SCREEN:

1.  On the **Home** screen, select the **Characterize your Fleets** button to display the **Fleet Characterization** screens.

2.  Confirm that the Fleet (Partner Name) that appears automatically is correct and appropriate for your fleets. If not, make changes in the field under the heading "Fleet (Partner Name)."

3.  Enter the "Fleet Identifier" for your first (or only) fleet. This field may be left blank if you only have one fleet. In this case your final Fleet Name will simply be your Partner Name.

4.  Enter additional fleets as needed:

    a.  To enter another fleet, select the **Add Another Fleet** button.

    b.  To delete a fleet, select the box next to the row you wish to delete, and then select the **Delete Checked Rows** button.

    c.  Once you have confirmed or modified the Partner Name and specified the Fleet Identifiers, the full Fleet Names will be displayed on the screen to the right, exactly how they will be displayed on the SmartWay website.

5.  To proceed, select the **Fleet Details** tab at the top, or simply select the **NEXT** button at the bottom of the screen.

6.  Before moving on, a popup screen will appear asking you to verify that you are satisfied with your fleet name(s).

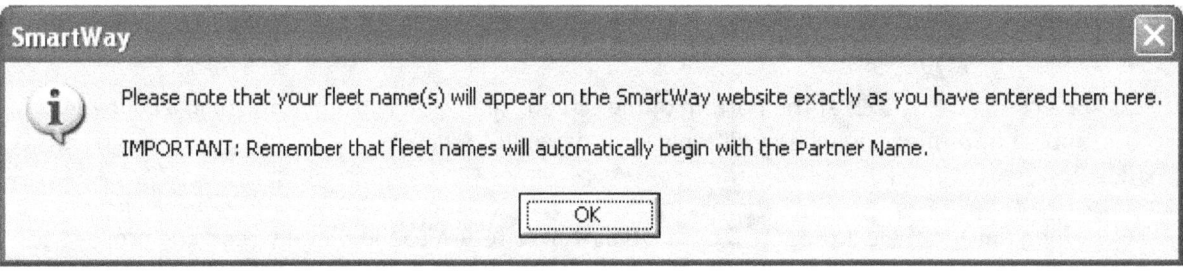

7.  ***Verify that the fleet name(s) show exactly what you want customers to find in the SmartWay website and Carrier Data File.***

8.  Select **OK** to proceed to the next screen. You may return to this screen later to revise your fleet name(s) if necessary.

## STEPS FOR COMPLETING "FLEET DETAILS" SCREEN

1. For each fleet, if you control over 95% of the operation of the vehicles (weighted by miles) check the box labeled "95+% Control."

2. Enter SCAC, MCN, and DOT identifiers (optional) for each fleet listed. If you have a single fleet that has multiple SCACs, enter all of them into the SCAC field, and separate them with commas.

   - While it is not required to enter SCAC, MCN, or DOT information for each fleet, it will help shippers and logistics companies searching by those parameters in the SmartWay database to easily find your fleet for inclusion in their Tool.

3. Select your Fleet Type (either "For-Hire" or "Private/Dedicated") for each fleet listed.

4. Select a Fleet Contact from the drop-down menu for each fleet listed. If the appropriate fleet contact is *not* listed, go back to the Contact Information screen (see Section 2), and enter that name under "Other Contacts." Then return to this screen to add the name from the drop-down menu.

5. Click the **NEXT** button or select the Operation Categories tab at the top to proceed to the next section.


## STEPS FOR COMPLETING "OPERATION CATEGORIES" SCREEN:

1. For each fleet identified, estimate the percentage of mileage each fleet spends in the five defined operation categories and enter your estimates in the fields provided. Leave the field blank if no mileage is associated with that operation category for that fleet.

2. Click the **NEXT** button or select the Body Types tab at the top to proceed to the next section.


## STEPS FOR COMPLETING "BODY TYPES" SCREEN:

1. For each fleet, enter an estimate of the percentage of fleet mileage associated with each body type. The percentages specified can be approximate, based on vehicle populations. The percentages for each fleet must sum to 100%.

**NOTE**: If you specify activity for Special Haulers, a **Describe** button will appear next to the cell entry. You must select this button and provide a text description of your specialty haulers.

2. Once you are sure your information is input correctly, you may select the **CREATE FLEET(S)** button at the bottom of the page. You will be prompted then to confirm that you have identified all of your fleets. Return to the Identify Fleets screen if needed, otherwise select **OK** and you will automatically be returned to the **Home** screen.

## STEPS FOR SELECTING A FLEET TO REVIEW:

1. Using your mouse, on the **Home** screen select and highlight the name of the fleet for which you wish to enter data.

2. Double click the name; you will then be taken to the General Information data entry screen for that fleet.

## STEPS FOR COMPLETING "GENERAL INFORMATION" SCREEN

1. Enter your percentage short-haul vs. long-haul operations.

   - Inputting a value in one cell automatically populates the other cell to add up to 100.

2. Check the boxes for the fuel types you use.

   - Once you check these boxes, the appropriate fuel type tab (along the top of the screen next to the General Information tab) will become active. If you select the **Diesel/Biodiesel** box, the grayed-out **Part 3: Particulate Matter Reduction** section will also become active.

3. Determine whether you will need to enter data for **Particulate Matter Reduction**; if yes, check the box.

   - Check the **Particulate Matter Reduction** box only if you have truck engines that are model year 2006 or earlier and are equipped with diesel retrofit particulate matter control devices (i.e., diesel oxidation catalysts (DOCs), particulate filters, or closed crankcase ventilation (CCV)).

4. Enter the percentage of trailers utilizing 100% of available cargo capacity while remaining within allowable weight limits.

5. Click the [ View/Select Commodities ] button and select which commodity categories you typically carry. Select all categories that apply to your fleets.

6. ***FOR FLEETS WITH 75% OR MORE OF TRUCKS IN THE DRAYAGE OPERATIONS CATEGORY:*** Check the box and enter the additional requested information on APUs and SmartWay tire use if you wish to participate in the Port Dray Program.

7. Select the [ VALIDATE SCREEN ] button to make sure you have filled out everything on this screen properly. You can also select [ VALIDATE FLEET ] to check your data entries across all screens for the given fleet.

8. Click [ NEXT ] or select the next available fuel tab at the top to begin entering fuel and activity data for this fleet.

## STEPS FOR COMPLETING THE "ENGINE MODEL YEAR & CLASS" SCREEN

***NOTE: The following guidance will use the Diesel Fuel sections as an example.
Similar procedures are followed for all other fuel types.***

Select the boxes at the top (i.e., 2b, 3, 4, 5, 6, 7, 8a, 8b) for each of the truck classes you operate in this fleet.

Input the number of vehicles you have in each class, specifying the corresponding **engine** years (rather than the tractor model years).

- Use the scroll bar to the right if you need to enter information for older model years.

Check the box at the bottom of the screen to hide any unused truck classes if you wish.

Select [NEXT] or select the Activity Information tab at top of the screen to proceed to the next section.

- Before leaving the Engine Model Year & Class screen, you will be prompted to confirm the accuracy of your model year and truck class selections. You may review previous years' selections by selecting the **Year-to-Year Comparison Report** on the Home screen.

## STEPS FOR COMPLETING THE FUEL DATA POINTS ON "ACTIVITY INFORMATION" SCREEN

1. Select the [Add] button under the **Overall Fleet & Data Source** section to specify where you obtained your data for each row. For each data type, specify the general type of data source and any additional data source detail information, as applicable.

2. Enter the exact value for total miles driven collectively by this fleet by vehicle class. Include all out-of-route, positioning, empty, and other miles driven.

3. Enter the exact value for the revenue miles—the number of miles your fleet drove that were charged to a customer account.

4. Enter the exact value for the total number of empty miles traveled by your fleet.

5. Enter the exact value for all the gallons of fuel used by your fleet in the **past 12-month reporting period**, including any gallons with biofuels (biodiesel for diesel vehicles, ethanol for gasoline vehicles).

6. If you specified any use of reefer body types under the Fleet Characterization section, the [Refrigeration Unit Fuel Consumption] button will appear. Select this button and provide your best estimate of your reefer fuel use in gallons for each truck class. Diesel, LPG, and electric trucks are assumed to use diesel reefers, gasoline trucks to use gasoline reefers, and CNG/LN G trucks to use CNG reefers. Enter zero for any truck classes/fuel types that do not utilize reefer units. If you do not know your reefer fuel use at the truck class level you may enter a total value which will be distributed across your truck classes proportional to your vehicle fuel consumption entries.

7. NOTE: On the Electric Vehicle Activity screen, "fuel inputs" are expressed in kWhrs rather than gallons.

## STEPS FOR COMPLETING THE "FUEL ALLOCATOR WORKSHEET"

1. Select the **Allocate Diesel Using class MPG** button on the Activity Information screen.
2. Enter total gallons of diesel for all truck classes.
3. Enter MPG estimates for each truck class.
4. If "Match" appears in blue next to the "Calculated Gallons Used" field, click **OK** .
5. If "No Match" appears in red next to the "Calculated Gallons Used" field, adjust the MPG values as necessary to get a match, and then click **OK** .

## STEPS FOR COMPLETING THE "BIODIESEL BLEND WORKSHEET "

1. If this fleet has used biodiesel, select the **Input Biodiesel** button and specify your biodiesel volumes by blend level in the Biodiesel Blend Worksheet.
2. For each of the blends used by your fleet, enter the appropriate number of gallons used.
3. Select **OK.**

## STEPS FOR COMPLETING THE "ETHANOL BLEND WORKSHEET "

1. If this fleet has used ethanol, select the **Input Ethanol** button and specify your ethanol volumes by blend level (E10 or E85) in the Ethanol Blend Worksheet.
2. For each of the blends used by your fleet, enter the appropriate number of gallons used.
3. Select **OK.**

## STEPS FOR COMPLETING THE PAYLOAD SECTION ON THE "ACTIVITY INFORMATION" SCREEN

1. Enter your average payloads into the Tool for each truck class by selecting the **Calc Payload** button.

2. Under the "Step 1: Getting Started Tab," select your activity Allocation Method.

3. Select your Units, from the drop-down.

4. Click the **Add** button next the "Data Source." A Data Source Description box will appear.

5. Using the drop-down menus, select your data source and details and enter any comments about your data source.

6. Click **OK** .

7. Click the "Step 2: Body Types" tab or **NEXT** .

8. Check the box(es) next to the body types used in this Truck Class.

9. For each body type selected, enter the activity associated with this body type.

10. For each body type selected, use the drop-down to select a range OR enter the exact payload if available. Make sure to include any pallet and packaging weight in your payload estimates.

11. Validate the screen to determine if there are any errors to correct; if yes, correct the errors and/or enter comments using the **ADD COMMENTS** button to explain your data inputs.

12. Click the **OK** button to perform a final validation check and return to the Activity Information screen.

## STEPS FOR COMPLETING AVERAGE CAPACITY VOLUME (CUBIC FEET) SECTION ON THE "ACTIVITY INFORMATION" SCREEN

For truck classes 8a and 8b follow these steps:

1. Click the **Calc Volume** button to open the Volume Calculator screen.

2. Click **Edit** to enter your data source for average capacity volume data.

3. Select one of the three reporting basis options using the radio check boxes.

4. Enter the requested metrics for the trailers, containers, tankers, bulk carrier, or other trailers used in this truck class.

5. Select **OK** to return to the Activity Information Screen.

For truck classes 2b through 7, follow these steps:

1. Click the **Calc Volume** button to open the Volume Calculator screen.

2. Under the "Step 1: Getting Started Tab," select your activity Allocation Method.

3. Select your Units, from the drop-down.

4. Click the **Add** button next the "Data Source." A Data Source Description box will appear.

5. Using the drop-down menus, select your data source and details and enter any comments about your data source.

6. Click **OK** .

7. Click the "Step 2: Body Types" tab or **NEXT** .

8. Check the box(es) next to the body types used in this Truck Class.

9. For each body type selected, enter the activity associated with this body type.

10. For each body type selected, use the drop-down to select the default average volume OR enter the exact volume if available.

11. Validate the screen to determine if there are any errors to correct; if yes, correct the errors and/or enter comments using the **ADD COMMENTS** button to explain your data inputs.

12. Click the **OK** button to perform a final validation check and return to the Activity Information screen.

## STEPS FOR COMPLETING THE % CAPACITY UTILIZATION SECTION ON THE "ACTIVITY INFORMATION" SCREEN

1. Next to the % Capacity Utilization (excluding empty miles) data entry box, click the [Add] button. A **Data Source Description** box will appear.

2. Using the drop-down menus, select your data source and details and enter any comments about your data source.

3. If you specified drayage activity under the **Operation Types** screen, you may select the data source "Unknown to dray carriers - use industry average" option, in which case the industry average capacity utilization value will be auto-populated for you.

4. Click [OK] to return to the **Activity Information** screen.

5. Enter your % capacity utilization for each truck class.

## STEPS FOR COMPLETING THE ROAD TYPE / SPEED CATEGORY SECTION ON THE "ACTIVITY INFORMATION" SCREEN

1. Click [Add] next to the data entry box next to "Road Type/Speed Categories." A **Data Source Description** box will appear.

2. Using the drop-down menus, select your data source and details and enter any comments about your data source.

3. Click [OK] to return to the **Activity Information** screen.

4. For each truck class, select the [Enter Speeds] button to open the ""Road Type/Speed Categories" box.

5. Enter the requested metrics for Highway or Rural Driving and Urban Driving for each of the speeds listed (if applicable). If you only know Highway or Rural driving, you can select the box next to "Populate the urban driving fields with default values" to complete the screen.

6. Click [OK] to return to the **Activity Information** screen.

## STEPS FOR COMPLETING THE AVERAGE ANNUAL IDLE HOURS PER TRUCK SECTION ON THE "ACTIVITY INFORMATION" SCREEN

1. Click [Add] next to the data entry box next to "Average Annual Idle Hours per Truck." A **Data Source Description** box will appear.

2. Using the drop-down menus, select your data source and details and enter any comments about your data source.

3. Click [OK] to return to the **Activity Information** Screen.

4. Click the  button to open the Idle Hours Calculator.

5. Enter daily long-duration idling hours per truck, daily short-duration idle hours per truck, and average days in service per year for each vehicle class represented in this fleet.

6. Click  to return to the Activity Information Screen.

---

<u>Next Steps</u>

You have now completed the **Activity Information** Screen for this fuel type.

Click the  button to verify that you have completed the screen properly.

**\*\*\* If you are using PM reduction equipment on model year 2006 or earlier trucks,** select the PM Reduction tab at the top of the screen to proceed to this section. Instructions for data entry are provided below this box. **\*\*\***

If you have finished entering data for this fuel type, select the tab for the next fuel type and complete all screens as indicated above.

If you have finished inputting data for all of your fuel types, select the [HOME] button to return to the Home screen.

<u>*REMEMBER:*</u>

\*\*\* If additional fuel types are represented in this fleet,
*you must complete the Engine Model Year & Class and Activity Information screens for <u>each</u> of the fuel types you operate.*

Select the tab for the next fuel type and complete the *Engine Model Year & Class and Activity Information screens* as indicated above—the steps will be the same or similar to those for the Diesel Tab. *Do NOT move on if you have not completed the data input for each of your fuel types.*\*\*\*

Once finished entering data for all of your fuel types, select the [HOME] button to return to the Home screen and follow the instructions to submit your Tool to EPA.

## STEPS FOR COMPLETING THE "PM REDUCTION" SCREEN

1. Select the radio button next to a device (DOC, CCV, or PM trap) you have used for your fleet.

2. Enter the number of trucks equipped with the device by <u>engine</u> (not vehicle) model year.

3. If other devices have been used with this fleet, select the radio box(es) next to each device and enter the number of trucks equipped with the device by <u>engine</u> (not vehicle) model year.

4. When done, select **VALIDATE SCREEN** to make sure you have filled out everything properly on this screen.

5. Select the **HOME** button to return to the Home screen.


## OPTIONAL STEPS—YEAR TO YEAR COMPARISONS, VIEW REPORTS, AND PROVIDE ADDITIONAL INFORMATION

After completing Steps 1 – 4 on the Home screen, you have access to three optional sections of the Tool.

The Year-to-Year Comparison Report allows the user to compare the fleet characteristics and activity values from your current reporting year, as well as $CO_2$ performance metrics, with your prior reporting year. This report is particularly helpful in identifying any changes that may have occurred since your last reporting period, determining trends in activity and performance, and performing general quality assurance of the inputs used for your current Tool. You can access this report by selecting **Review Year-to-Year Comparison** under item #5 on the Home screen.

The Partner Profile/Logo Info/Suggestions section allows you to provide EPA with additional information about your company, potential use of the SmartWay Logos, and general feedback regarding the SmartWay program. This information is optional and is not required in order to submit your Truck Carrier Tool data to EPA. Selecting the **Partner Profile / Logo Info / Suggestions** button under item #6 on the Home screen will open the Partner Information forms.

The **View Your Data Reports** section gives you access to 13 reports to help you understand your data and use it to make better performance decisions. You can access these reports by selecting **View Your Data Reports** under item #7 on the Home screen.

*\*\*\*Once you've reviewed these sections, you will be ready to submit your Tool to EPA.\*\*\**

## STEPS FOR SUBMITTING YOUR TOOL TO EPA

1. Select the | Generate File to Send to EPA* | button, which will open a new screen.

2. Select the checkbox to indicate you understand the terms of the SmartWay Partnership Agreement once again.

3. Next, a question will appear asking if you are an existing SmartWay Partner (Y/N). If you are, a question will appear asking if you submitted your data the previous reporting year.

   a. If so, you must then enter your Annual Submission ID, which has been sent to you by your SmartWay PAM via email. The SmartWay ID is 8 digits in length.

   b. If you cannot locate your submission ID you can select the link "Email me my annual submission ID" to have your ID sent to you.

4. When ready, select NEXT to create a file with the following naming convention:

   Truck_PartnerName_ Year_ V0.xml[10]

   For example, Truck_ABCompany_ 2013 _V0.xml

where PartnerName is your company's name as entered for Step 1 on the Home screen, and Year indicates the year for which you are submitting your data.

5. Next specify the folder where you would like to save the xml file, and a screen will appear.

6. Follow these instructions for submitting your xml file to SmartWay. Note that the .xml file is approximately 10 times smaller than the .xls file.

7. Upon selecting NEXT, a screen will appear that allows you to close the Truck Carrier Tool or return to the Home screen.

   - **NOTE**: DO NOT ZIP the File. Send it to EPA as a normal file attached in an e-mail. EPA security will not allow zipped files through the EPA firewall.

   - **NOTE**: DO NOT CHANGE THE NAME OF THE XML FILE.

Please visit http://www.epa.gov/smartway/partnership/trucks.htm for additional resources, including printable data collection worksheets, technical documentation, and the **Truck Carrier Tool Data Entry and Troubleshooting Guide** for more in depth exploration of each section of the Truck Carrier Tool.

---

[10] If you create the XML file multiple times the file name will increment each time (e.g., Truck_ABCompany_2013_V1.XML for the second iteration, etc.

# APPENDIX A: RECOMMENDED DATA SOURCES FOR ACTIVITY DATA

**Table 1** summarizes **the standard Data Source categories** available for selection for each data type.

### Table 1: Data Source Detail Selection Options

| Data Type | Data Source | Data Source Detail |
|---|---|---|
| Total Miles Driven | As reported to IFTA Form 441 for tax reporting (interstate) – Class 7, 8a and 8b trucks only | Collected via fleet-wide GPS reporting software |
| | | Collected via odometer readings |
| | | Collected via maintenance records |
| | | Collected via driver trip sheets |
| | | Collected via standard mileage routes, e.g. PC Miler, Household Goods Guide |
| | As reported to individual states for tax reporting (intrastate) | Collected via fleet-wide GPS reporting software |
| | | Collected via odometer readings |
| | | Collected via maintenance records |
| | | Collected via driver trip sheets |
| | | Collected via standard mileage routes, e.g. PC Miler, Household Goods Guide |
| | Determined using software | Dispatching Software* |
| | | Transportation Management System (TMS)* |
| | Vehicle-based data collection | Determined via Electronic Control Module (ECM) data recorder/logger* |
| Revenue Miles Driven | As used in Federal tax reporting | Collected via electronic Transportation Management System (TMS) |
| | | Collected via GPS-enabled TMS |
| | | Collected via manual input into company database with driver trip sheets |
| | | Collected via non-electronic company records |
| | Financial data | Accounting/billing software* |
| | | Tax reports/IRS and State* |

| Data Type | Data Source | Data Source Detail |
|---|---|---|
| **Revenue Miles Driven (cont'd)** | Determined using software | Dispatching Software* |
| | | Transportation Management System (TMS)* |
| | Based on total mileage | Equal to total miles |
| | | Total miles less empty miles |
| | | Calculated as a percentage of total miles* |
| **Empty Miles Driven** | Collected automatically / electronically / manually | Collected via electronic Transportation Management System (TMS) |
| | | Collected via GPS-enabled TMS |
| | | Collected via manual input into company database |
| | | Collected via non-electronic company records |
| | | Collected via odometer readings |
| | | Collected via driver trip sheets |
| | Financial Data | Accounting/billing software* |
| | | Tax reports/IRS and State* |
| | Determined using software | Dispatching Software* |
| | | Transportation Management System (TMS)* |
| | Based on total mileage | Total miles less revenue miles |
| | | Calculated as a percentage of total miles* |
| **Gallons of Fuel Used^** | As reported to IFTA Form 441 for tax reporting (interstate) – Class 7, 8a and 8b trucks only | Collected via electronic fuel receipt |
| | | Collected via paper fuel receipt |
| | | Collected via driver trip sheets |
| | | Collected via electronic expenditure data |
| | | Collected via paper expenditure data |
| | As reported to individual state for tax reporting (intrastate) | Collected via electronic fuel receipt |
| | | Collected via paper fuel receipt |
| | | Collected via driver trip sheets |
| | | Collected via electronic expenditure data |

| Data Type | Data Source | Data Source Detail |
|---|---|---|
| **Gallons of Fuel Used (cont'd)** | | Collected via paper expenditure data |
| | Determined using software | Dispatching Software* |
| | | Transportation Management System (TMS)* |
| | Vehicle-based data collection | Determined via Electronic Control Module (ECM) data recorder/logger* |
| | Based on MPG estimates* | User-provided |
| **Average Payload** | Bills of Lading – electronic records (preferred) | Based on actual miles traveled by specific payloads* |
| | | Trip-weighted (total payload weights / total trips)* |
| | Bills of Lading – manual records | Based on actual miles traveled by specific payloads* |
| | | Trip-weighted (total payload weights / total trips)* |
| | Ranges provided by calculator | N/A (calculator) |
| **Average Volume** | Determined using company records* | User-provided |
| | Defaults from calculator | N/A (calculator) |
| **Capacity Utilization** | Collected automatically / electronically / manually | Collected through load volume information |
| | Determined using software | Dispatching Software* |
| | | Transportation Management System (TMS)* |
| **Road Type / Speed Categories** | Collected automatically / electronically | Driver trip sheets* |
| | | Governed speed* |
| | | Determined via GPS |
| | | Determined via Electronic Control Module (ECM) data recorder/logger* |
| | Transportation Management System (TMS) | Driver trip sheets* |
| | | Governed speed* |
| | | Determined via GPS |
| | | Determined via Electronic Control Module (ECM) data recorder/logger* |

| Data Type | Data Source | Data Source Detail |
|---|---|---|
| **Average Annual Idle Hours per Truck** | Vehicle-based data collection | Determined via Electronic Control Module (ECM) data recorder/logger* |
| | Driver trip reports* | User-provided |
| | Idle reduction strategy | Company "No Idle" policy in place* |
| | | Local/State idle regulation in place* |
| | Determined using software | Dispatching Software* |
| | | Transportation Management System (TMS)* |

\* User must provide additional description regarding data collection system and calculation method.

^ For electric vehicles, in lieu of gallons, use kWhrs. Common data sources for kWhrs include metering at charging dock, vehicle data acquisition units, and smart-meter applications.

# APPENDIX B: WORKSHEETS FOR DATA COLLECTION

**List of Worksheets**

## Worksheet #1: US/Canada Operation Information

% of fleet(s) licensed in the United States:____

% of fleet(s) licensed in the Canada:      ____

% of fuel consumed by US licensed vehicles: ____

% of fuel consumed by Canada licensed vehicles: ____

Data Source used for determining % of fuel consumed (circle one):

- **As reported to IFTA Form 441 for tax reporting (interstate)** – Class 7, 8a and 8b trucks only
- **As reported to individual state for tax reporting (intrastate)**
- **Determined using software**
- **Based on MPG estimates**
- **Other – provide description:**

  _____

Do your US and Canadian-licensed vehicles generally operate using the same fuel mix?  (Circle one)

- Yes
- No
- N/A (only operate in one country)

## WORKSHEET #2: CONTACT INFORMATION

#1. Enter your Contact Information:

### General Company Contact Information

| Company Name | | | | | | | | |
|---|---|---|---|---|---|---|---|---|
| Headquarters Mailing Address | | | | | | | | |
| City | | | State/Province | | Zip | | Country | |
| Main Phone Number | | Toll-free Number | | Cell number | | Web Address | | |

### Primary Contact Information

| Primary Contact Name | | | | | | | |
|---|---|---|---|---|---|---|---|
| Primary Contact Mailing Address | | | | | | | |
| City | | State/Province | | Zip | | Country | |
| Primary Contact Phone Number | | | Email Address | | | | |

### Executive Contact Information

| Executive Contact Name | | | | | | | |
|---|---|---|---|---|---|---|---|
| Executive Contact Mailing Address | | | | | | | |
| City | | State/Province | | Zip | | Country | |
| Executive Contact Phone Number | | | Email Address | | | | |

### Other Contact Information

| Executive Contact Name | | | | | | | |
|---|---|---|---|---|---|---|---|
| Executive Contact Mailing Address | | | | | | | |
| City | | State/Province | | Zip | | Country | |
| Executive Contact Phone Number | | | Email Address | | | | |
| Contact's role in program | | | | | | | |

## WORKSHEET #3: FLEET CHARACTERIZATION

Complete this worksheet for <u>each fleet</u> you will be submitting in the Truck Carrier Tool.

Specify Fleet Name:  This will be a combination of your Partner Name and a Fleet Identifier you create. Use a Fleet Identifier that a company hiring your fleet would recognize. Enter it exactly as it should appear on the SmartWay website):

Partner Name _____

Fleet Identifier (suffix)_____

95% Control _____  SCAC: _____  MCN:_____  DOT #: _____FLEET TYPE:_____

Fleet Contact:_____

**Operation Category Percentages:**
Truckload _____  LTL _____  Drayage _____  Package Delivery_____  Expedited _____

**Body Type Percentages:**
Dry Van_____  Reefer _____  Flatbed_____  Tanker_____  Chassis _____
Heavy-Bulk _____  Auto Carrier _____  Moving ____  Utility _____  Special Hauler_____


**If Special Hauler is selected, please describe type:**

_____

## WORKSHEET #4: GENERAL FLEET INFORMATION (PAGE 1 OF 2)

**Long- versus Short-haul split (%):** _____ Short Haul _____ Long Haul

**Types of Fuel Used:**

_____ diesel/biodiesel _____ gasoline/ethanol _____ propane (LPG)

_____ liquefied natural gas (LNG) _____ compressed natural gas (CNG)

_____ electric _____ hybrid electric (diesel/gasoline)

**Use of Particulate Matter (PM) Control Equipment:**

_____ Diesel Oxidation Catalyst (DOC)

_____ Closed Crankcase Ventilation (CCV)

_____ Particulate Matter Filter (PM Trap )

**Cube Out Percentage:** _____

| | Commodity Descriptions: Select all that apply to this fleet: |
|---|---|
| | Animals and Fish (Live) |
| | Cereal Grains (including seed) |
| | Agricultural Products except for Animal Feed |
| | Animal Feed and Products of Animal Origin |
| | Meat, Fish, and Seafood, and their Preparations |
| | Milled Grain Products and Preparations, and Bakery Products |
| | Other Prepared Foodstuffs, and Fats and Oils |
| | Alcoholic Beverages and Tobacco Products |
| | Stone and Sands, Except Metal Bearing Sands |
| | Other Non-Metallic Minerals |
| | Metallic Ores and Concentrates |
| | Coal |
| | Crude Petroleum, Gasoline, Fuel Oils, and Aviation Turbine Fuel |
| | Other Coal and Petroleum Products |
| | Basic Chemicals |
| | Pharmaceutical Products |
| | Fertilizers |
| | Other Chemical Products and Preparations |
| | Plastics and Rubber |
| | Logs and Other Wood in the Rough |
| | Wood Products |
| | Pulp, Newsprint, Paper, and Paperboard |
| | Paper or Paperboard Articles |
| | Printed Products |

| | |
|---|---|
| | Textiles, Leather, and Articles of Textiles or Leather |
| | Non-Metallic Mineral Products |
| | Base Metal in Primary or Semi-Finished Forms and in Finished Basic Shapes |
| | Articles of Base Metal |
| | Machinery |
| | Electronic and Other Electrical Equipment and Components, and Office Equipment |
| | Motorized and Other Vehicles (including parts) |
| | Transportation Equipment |
| | Precision Instruments and Apparatus |
| | Furniture, Mattresses and Mattress Supports, Lamps, Lighting Fittings, and Illuminated Signs |
| | Miscellaneous Manufactured Products |
| | Waste and Scrap (except for agriculture or food) |
| | Mixed Freight |

## WORKSHEET #5A MODEL YEAR/CLASS

NOTE: Print multiple copies of this worksheet to gather data for each fuel type within each fleet.

**Fleet Name:** _____

**Fuel Type:** _____

**Engine Model Year(s) and Class(es):**

| | Class 2b 8,501-10,000 | Class 3 10,001-14,000 | Class 4 14,001-16,000 | Class 5 16,001-19,500 | Class 6 19,501-26,000 | Class 7 26,001-33,000 | Class 8a 33,001-60,000 | Class 8b 60,001 & above |
|---|---|---|---|---|---|---|---|---|
| **ENGINE MODEL YEAR** | 2B | 3 | 4 | 5 | 6 | 7 | 8A | 8B |
| 2014 | | | | | | | | |
| 2013 | | | | | | | | |
| 2012 | | | | | | | | |
| 2011 | | | | | | | | |
| 2010 | | | | | | | | |
| 2009 | | | | | | | | |
| 2008 | | | | | | | | |

| | Class 2b 8,501-10,000 | Class 3 10,001-14,000 | Class 4 14,001-16,000 | Class 5 16,001-19,500 | Class 6 19,501-26,000 | Class 7 26,001-33,000 | Class 8a 33,001-60,000 | Class 8b 60,001 & above |
|---|---|---|---|---|---|---|---|---|
| ENGINE MODEL YEAR | 2B | 3 | 4 | 5 | 6 | 7 | 8A | 8B |
| 2007 | | | | | | | | |
| 2006 | | | | | | | | |
| 2005 | | | | | | | | |
| 2004 | | | | | | | | |
| 2003 | | | | | | | | |
| 2002 | | | | | | | | |
| 2001 | | | | | | | | |
| 2000 | | | | | | | | |
| 1999 | | | | | | | | |
| 1998 | | | | | | | | |
| 1997 | | | | | | | | |
| 1996 | | | | | | | | |

| ENGINE MODEL YEAR | Class 2b 8,501-10,000 2B | Class 3 10,001-14,000 3 | Class 4 14,001-16,000 4 | Class 5 16,001-19,500 5 | Class 6 19,501-26,000 6 | Class 7 26,001-33,000 7 | Class 8a 33,001-60,000 8A | Class 8b 60,001 & above 8B |
|---|---|---|---|---|---|---|---|---|
| 1995 | | | | | | | | |
| 1994 | | | | | | | | |
| 1993 | | | | | | | | |
| 1992 | | | | | | | | |
| 1991 | | | | | | | | |
| 1990 | | | | | | | | |
| 1989 | | | | | | | | |
| 1988 | | | | | | | | |
| Pre-1988 | | | | | | | | |

# WORKSHEET #5B: ACTIVITY DATA SOURCES AND INFORMATION (PAGE 1 OF 2)

## Fleet -Level Data Sources:

Total Miles Driven _____

Revenue Miles Driven _____

Empty Miles Driven _____

Gallons of Fuel Used (incl. Biofuel – biodiesel or ethanol) _____

Gallons of Reefer Fuel Used (diesel, gasoline, CNG - by vehicle class) _____

% Capacity Utilization (excluding empty miles) _____

Road Type / Speed Category _____

Average Idle Hours per Truck _____

## Truck Class-Level Data Sources (payload and volume only)

| | 2B | 3 | 4 | 5 |
|---|---|---|---|---|
| Average Payload (tons or pounds – circle one) Cargo Weight Only | | | | |
| Average Capacity Volume (cubic feet or TEUs – circle one) | | | | |

| | 6 | 7 | 8A | 8B |
|---|---|---|---|---|
| Average Payload (tons or pounds – circle one) Cargo Weight Only | | | | |
| Average Capacity Volume (cubic feet or TEUs – circle one) | | | | |

# WORKSHEET #5B: ACTIVITY DATA SOURCES AND INFORMATION (PAGE 2 OF 2)

## Activity Data by Truck Class

| | 2B | 3 | 4 | 5 | 6 | 7 | 8A | 8B |
|---|---|---|---|---|---|---|---|---|
| Total Miles Driven (exact values) | | | | | | | | |
| Revenue Miles Driven (exact values) | | | | | | | | |
| Empty Miles Driven (exact values) | | | | | | | | |
| Gallons of Fuel used, including biofuels and Reefer used (exact values) | | | | | | | | |
| % Capacity Utilization (excluding empty miles) | | | | | | | | |

# WORKSHEET #5C: ACTIVITY - BIODIESEL AND ETHANOL BLENDS (IF APPLICABLE)

For each of the biodiesel blends used by your fleet, note the appropriate number of gallons used.

| | | | | | | | | | |
|---|---|---|---|---|---|---|---|---|---|
| B1 | | B21 | | B41 | | B61 | | B81 | |
| B2 | | B22 | | B42 | | B62 | | B82 | |
| B3 | | B23 | | B43 | | B63 | | B83 | |
| B4 | | B24 | | B44 | | B64 | | B84 | |
| B5 | | B25 | | B45 | | B65 | | B85 | |
| B6 | | B26 | | B46 | | B66 | | B86 | |
| B7 | | B27 | | B47 | | B67 | | B87 | |
| B8 | | B28 | | B48 | | B68 | | B88 | |
| B9 | | B29 | | B49 | | B69 | | B89 | |
| B10 | | B30 | | B50 | | B70 | | B90 | |
| B11 | | B31 | | B51 | | B71 | | B91 | |
| B12 | | B32 | | B52 | | B72 | | B92 | |
| B13 | | B33 | | B53 | | B73 | | B93 | |
| B14 | | B34 | | B54 | | B74 | | B94 | |
| B15 | | B35 | | B55 | | B75 | | B95 | |
| B16 | | B36 | | B56 | | B76 | | B96 | |
| B17 | | B37 | | B57 | | B77 | | B97 | |
| B18 | | B38 | | B58 | | B78 | | B98 | |
| B19 | | B39 | | B59 | | B79 | | B99 | |
| B20 | | B40 | | B60 | | B80 | | B100 | |

| Actual ethanol usage figures: | Blend percentage: | Total Gallons Used |
|---|---|---|
| | E0 | |
| | E10 | |
| | E85 | |

To determine average payload for the entire fleet you will need to select an allocation method to determine the amount each truck class/vehicle type contributes to the overall total. Four allocation methods are provided, listed from most preferred to least preferred. You will also need to specify the data source(s) used to develop your estimates, and select the units for Payload (short tons or pounds) and Volume (cubic feet or TEU).

# CLASS 2B—WORKSHEET #5D: PAYLOAD AND VOLUME (PAGE 2 OF 12)

| Select allocation method for this class (circle one): | # of miles | # of trips | % of operation | # of vehicles |
|---|---|---|---|---|
| | | | | |

Select payload unit (circle one):   Tons    Pounds

Select volume unit (circle one):   Cubic Ft    TEU

| Vehicle Class | Vehicle Type | Allocation factor (#miles, trips, etc.) | Average Payload | Describe Payload Data Source/Calculation Method | Average Volume | Describe Volume Data Source/Calculation Method |
|---|---|---|---|---|---|---|
| 2B | Flatbed | | | | | |
| 2B | Step van | | | | | |
| 2B | Walk In | | | | | |
| 2B | Conventional van | | | | | |
| 2B | ? OTHER | | | | | |

## CLASS 3—WORKSHEET #5D: PAYLOAD AND VOLUME (PAGE 3 OF 12)

| Select allocation method for this class (circle one): | # of miles | # of trips | % of operation | # of vehicles |
|---|---|---|---|---|
| | | | | |

Select payload unit (circle one):     Tons     Pounds

Select volume unit (circle one):     Cubic Ft     TEU

| Vehicle Class | Vehicle Type | Allocation factor (#miles, trips, etc.) | Average Payload | Describe Payload Data Source/Calculation Method | Average Volume | Describe Volume Data Source/Calculation Method |
|---|---|---|---|---|---|---|
| 3 | Step van | | | | | |
| 3 | Walk In | | | | | |
| 3 | Conventional van | | | | | |
| 3 | **?** OTHER | | | | | |

# CLASS 4—WORKSHEET #5D: PAYLOAD AND VOLUME (PAGE 4 OF 12)

| Select allocation method for this class (circle one): | # of miles | # of trips | % of operation | # of vehicles |
|---|---|---|---|---|
| | | | | |

Select payload unit (circle one):　　Tons　　Pounds

Select volume unit (circle one):　　Cubic Ft　　TEU

| Vehicle Class | Vehicle Type | Allocation factor (#miles, trips, etc.) | Average Payload | Describe Payload Data Source/Calculation Method | Average Volume | Describe Volume Data Source/Calculation Method |
|---|---|---|---|---|---|---|
| 4 | Flatbed | | | | | |
| 4 | Step van | | | | | |
| 4 | Large walk-in | | | | | |
| 4 | Conventional van | | | | | |
| 4 | ? OTHER | | | | | |

# CLASS 5—WORKSHEET #5D: PAYLOAD AND VOLUME (PAGE 5 OF 12)

| Select allocation method for this class (circle one): | # of miles | # of trips | % of operation | # of vehicles |
|---|---|---|---|---|
| | | | | |

Select payload unit (circle one):   Tons   Pounds

Select volume unit (circle one):   Cubic Ft   TEU

| Vehicle Class | Vehicle Type | Allocation factor (#miles, trips, etc.) | Average Payload | Describe Payload Data Source/Calculation Method | Average Volume | Describe Volume Data Source/Calculation Method |
|---|---|---|---|---|---|---|
| 5 | Large walk-in | | | | | |
| 5 | Conventional van | | | | | |
| 5 | ? OTHER | | | | | |

# CLASS 6—WORKSHEET #5D: PAYLOAD AND VOLUME (PAGE 6 OF 12)

| Select allocation method for this class (circle one): | # of miles | # of trips | % of operation | # of vehicles |
|---|---|---|---|---|
| | | | | |

Select payload unit (circle one):  Tons  Pounds

Select volume unit (circle one):  Cubic Ft  TEU

| Vehicle Class | Vehicle Type | Allocation factor (#miles, trips, etc.) | Average Payload | Describe Payload Data Source/Calculation Method | Average Volume | Describe Volume Data Source/Calculation Method |
|---|---|---|---|---|---|---|
| 6 | Flatbed | | | | | |
| 6 | Reefer | | | | | |
| 6 | Large walk-in | | | | | |
| 6 | Single axle van | | | | | |
| 6 | ? OTHER | | | | | |

# CLASS 7 STRAIGHT TRUCKS—WORKSHEET #5D: PAYLOAD AND VOLUME (PAGE 7 OF 12)

| Select allocation method for this class (circle one): | # of miles | # of trips | % of operation | # of vehicles |
|---|---|---|---|---|
| | | | | |

Select payload unit (circle one):　　Tons　　Pounds

Select volume unit (circle one):　　Cubic Ft　　TEU

| Vehicle Class | Vehicle Type | Allocation factor (#miles, trips, etc.) | Average Payload | Describe Payload Data Source/Calculation Method | Average Volume | Describe Volume Data Source/Calculation Method |
|---|---|---|---|---|---|---|
| 7 | Beverage | | | | | |
| 7 | Flatbed | | | | | |
| 7 | Reefer | | | | | |
| 7 | Tanker | | | | | |
| 7 | Single axle van | | | | | |
| 7 | ? OTHER | | | | | |

| Select allocation method for this class (circle one): | # of miles | # of trips | % of operation | # of vehicles |
|---|---|---|---|---|
| | | | | |

Select payload unit (circle one):     Tons          Pounds

Select volume unit (circle one):     Cubic Ft     TEU

| Vehicle Class | Vehicle Type | Allocation factor (#miles, trips, etc.) | Average Payload | Describe Payload Data Source/Calculation Method | Average Volume | Describe Volume Data Source/Calculation Method |
|---|---|---|---|---|---|---|
| 7 | Combination Flatbed | | | | | |
| 7 | Combination Reefer | | | | | |
| 7 | Dry Van – Single | | | | | |
| 7 | OTHER | | | | | |

| Select allocation method for this class (circle one): | # of miles | # of trips | % of operation | # of vehicles |
|---|---|---|---|---|
| | | | | |

Select payload unit (circle one):     Tons          Pounds

| Vehicle Class | Vehicle Type | Allocation factor (#miles, trips, etc.) | Average Payload | Describe Payload Data Source/Calculation Method |
|---|---|---|---|---|
| 8A | Flatbed | | | |
| 8A | Tanker | | | |
| 8A | Single axle van | | | |
| 8A | **?** OTHER | | | |

| Select allocation method for this class (circle one): | # of miles | # of trips | % of operation | # of vehicles |
|---|---|---|---|---|
| | | | | |

Select payload unit (circle one):     Tons          Pounds

| Vehicle Class | Vehicle Type | Allocation factor (#miles, trips, etc.) | Average Payload | Describe Payload Data Source/Calculation Method |
|---|---|---|---|---|
| 8A | Beverage | | | |
| 8A | Combination Flatbed | | | |
| 8A | Dry Van - Single | | | |
| 8A | ? OTHER | | | |

| Select allocation method for this class (circle one): | # of miles | # of trips | % of operation | # of vehicles |
|---|---|---|---|---|
| | | | | |

Select payload unit (circle one):     Tons          Pounds

| Vehicle Class | Vehicle Type | Allocation factor (#miles, trips, etc.) | Average Payload | Describe Payload Data Source/Calculation Method |
|---|---|---|---|---|
| 8B | Dry Van - Single | | | |
| 8B | Dry Van - Double | | | |
| 8B | Dry Van - Triple | | | |
| 8B | Combination Reefer | | | |
| 8B | Combination Flatbed | | | |
| 8B | Combination Tanker | | | |
| 8B | Chassis | | | |
| 8B | Specialty | | | |
| 8B | ? OTHER | | | |

## _Volume—Worksheet #5D: Payload and Volume (page 12 of 12)_

The volume worksheet for Class 8A/B reflects a range of standard trailer and container configurations. This worksheet will help you prepare for completing the worksheet in the Truck Carrier Tool.

Please specify your data source:

_____

_____

Please specify your reporting basis:

___Percent Usage      ___Number of Truckloads/Year   ___Number of Trailers

Enter the numbers by type of trailer or container; the Tool will calculate an average capacity volume for your fleet:

| Trailer Size | Percentage/# Truckloads/# Trailers | Container Size | Percentage/# Truckloads/# Trailers | Other Trailers | |
|---|---|---|---|---|---|
| 20 ft | | 20 ft | | Cubic Feet per Truck | |
| 40 ft | | 40 ft | | Number/Percent | |
| 42 ft | | 45 ft | | | |
| 45 ft | | 48 ft | | | |
| 48 ft | | 53 ft | | | |
| 53 ft | | | | | |
| 57 ft | | **Bulk Carrier Size** | | | |
| 28 + 28 | | Large (42" X 8.5" X 11.5") | | | |
| 40 + 40 | | Medium (32" X 8" X 11") | | | |
| 48 + 28 | | Small (22" X 8" X 10.25") | | | |
| 48 + 48 | | | | | |
| 28 + 28 + 28 | | **Liquid Tanker Size** | | | |
| | | Large (7,500+ gallons) | | | |
| | | Medium (3,001-7,499 gallons) | | | |
| | | Small (3,000 and under gallons) | | | |

*WORKSHEET #5E: ROAD TYPE/SPEED DISTRIBUTION AND IDLE HOURS WORKSHEET*

| | 2B | 3 | 4 | 5 | 6 | 7 | 8B | 8B | 8B |
|---|---|---|---|---|---|---|---|---|---|
| **ROAD TYPE / SPEED DISTRIBUTION** | | | | | | | | | |
| % Highway vs. Rural Driving | | | | | | | | | |
| % Urban Driving Under 25 mph | | | | | | | | | |
| % Urban Driving 25 to 50 mph | | | | | | | | | |
| % Urban Driving 50 mph+ | | | | | | | | | |
| **AVERAGE DAILY IDLE HOURS PER TRUCK** | | | | | | | | | |
| Daily Long Duration Idle Hours Per Truck | | | | | | | | | |
| Daily Short Duration Idle Hours Per Truck | | | | | | | | | |
| Average Days in Service Per Truck Per Year | | | | | | | | | |

# WORKSHEET #5F: DIESEL PM WORKSHEET

**Trucks with PM Controls by truck class – Note # of trucks by control type (CCV, DOC, PM Trap, Flow-through Filter)**

| ENGINE MODEL YEAR | 2B | 3 | 4 | 5 | 6 | 7 | 8B | 8B |
|---|---|---|---|---|---|---|---|---|
| 2006 | | | | | | | | |
| 2005 | | | | | | | | |
| 2004 | | | | | | | | |
| 2003 | | | | | | | | |
| 2002 | | | | | | | | |
| 2001 | | | | | | | | |
| 2000 | | | | | | | | |
| 1999 | | | | | | | | |
| 1998 | | | | | | | | |
| 1997 | | | | | | | | |
| 1996 | | | | | | | | |
| 1995 | | | | | | | | |
| 1994 | | | | | | | | |
| 1993 | | | | | | | | |
| 1992 | | | | | | | | |
| 1991 | | | | | | | | |
| 1990 | | | | | | | | |
| 1989 | | | | | | | | |
| 1988 | | | | | | | | |
| Pre-1988 | | | | | | | | |

## WORKSHEET #5G: PORT DRAYAGE PROGRAM DATA REQUIREMENTS

Provide the following additional information for each fleet participating in the SmartWay Port Drayage Program.

**# of trucks equipped with auxiliary power units (APUs): _____**

**# of trucks equipped with SmartWay tires: _____**